Intellectual Property Lawyers

Joy Yongo

Series Editor **Casey Malarcher**

Level 4 - ⑤

Intellectual Property Lawyers

Joy Yongo

Series Editor: Casey Malarcher
Acquisitions Editor: Anne Taylor
Copy Editor: Liana Robinson
Cover/Interior Design: Highline Studio

ISBN: 978-1-943980-52-9

10 9 8 7 6 5 4 3 2 1
22 21 20 19 18

Photo Credits

Contents

What Is Intellectual Property?

Property can be anything that we own. It can be something that we buy at the store like a bike. It can also be something big such as a car or a house. Intellectual property is something that we own, too. But usually it cannot be held. It is what we create in our minds.

Creative ▶
thinking

A book full of ideas

One type of intellectual property is a design. At any McDonald's, you will see a big, yellow M. The design of the M is a logo. It is protected by a trademark. That means no one else can use the same design or logo as McDonald's.

A design protected by a trademark

Similarly, a brand name is intellectual property, too. A brand name helps people know what products or services are related to a company. A company such as Google is a powerful brand. Everyone knows Google. They trust Google and its products. Google is also trademarked. So, no one else can use the same name to do business.

◀ The company's brand name

A famous company

Another type of intellectual property is an invention. When someone creates a new object, they get a patent to protect it. This means that no one else can make the exact same object and sell it for money. When Apple made the iPhone, they received a patent. No one else is allowed to sell the exact same phone for money. It is against the law.

Patents protect ▶
inventions.

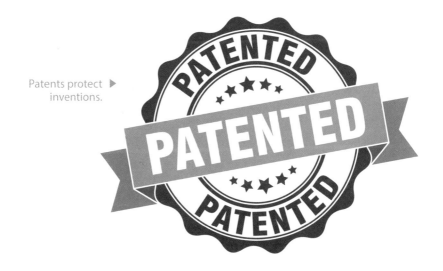

Agreeing to work together

Trade secrets are similar to inventions. A company's trade secret is what makes its product different and special. If it is a food product, the trade secret might be a secret ingredient. In contrast to inventions, trade secrets are usually protected by agreements. Only a few people in the company know the secret. They agree to keep the secret to themselves. Trade secrets are also protected by intellectual property laws.

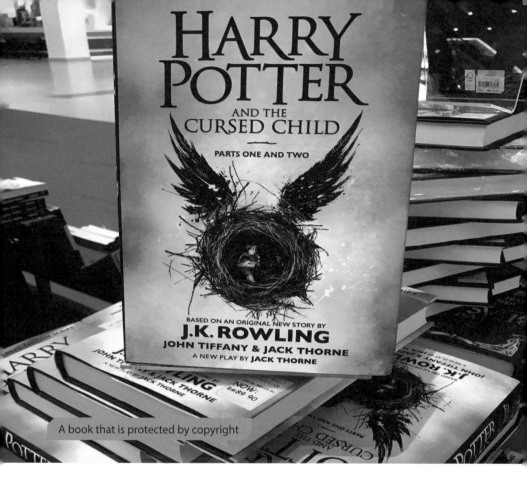

Things that artists create are intellectual property, too. These are original works. They could be songs, pieces of art, or books. Artistic works also include photographs and building designs. These are all protected by copyrights. No one else can use them in any way to sell them or make money from them without permission.

Text from a ▶
copyright page

◀ The copyright
symbol

The Job of an Intellectual Property Lawyer

Intellectual property lawyers have two main jobs. First, they do everything that is needed to protect intellectual property. There are many rules for patents, trademarks, and copyrights. The process to get one of these can be confusing and difficult. People can apply for one on their own. But it is much better to have the help of an intellectual property lawyer.

Lawyers write and review agreements

Creating a contract

A good intellectual property lawyer will know the rules and processes well. Intellectual property lawyers can also help write agreements for licensing works with copyrights or trade secrets of companies. They can also help write other types of contracts.

The wording for any application, agreement, or contract should be exact. It should be specific. In fact, the more details it has, the better. Why? Because it helps protect the intellectual property.

When details are left out or are not clearly explained, people may try to find loopholes in the agreement. If they do, they can steal intellectual property without actually breaking the law. This would be very bad. That is why people need good intellectual property lawyers.

◀ Tearing up a bad contract

Finding a problem

Sometimes, someone may try to use or sell something without the permission of the company or the patent owner. In these cases, intellectual property lawyers will help defend the patent. If the company or patent owner wins, they will receive money for any loss of profits. In addition, if another company accuses you of interfering with their patent, an intellectual property lawyer can represent you in court.

A disagreement ▶

Defending a case in court

14

A very famous patent lawsuit happened between Apple and Samsung. In 2011, Apple sued Samsung. It claimed that Samsung copied certain parts of its iPhone design. Then Samsung decided to sue Apple back. It claimed that Apple interfered with its technology patents.

Phones made by Samsung and Apple

Apple's brand

Samsung's brand

Both companies together were selling more than half of the world's smartphones at that point. Who interfered with whose patents? Who should pay who? What was the damage? The patent war was a war of billions of dollars. And at the front of it all were intellectual property lawyers representing each company.

A court case ▶
with lots of
money at stake

How to Be an Intellectual Property Lawyer

Becoming an intellectual property lawyer is not an easy thing to do. But it's not impossible. You need to get a university degree. It can be a science or math degree. It can be a degree in history, social science, or art. The choice is up to you. But a degree is not the end. It is only the beginning.

University students

A lecture hall

After finishing university, you will need to take a special test. Anyone who wants to go to law school must take this kind of test. In graduate school, every law student will learn the basics of law. A lawyer must learn how to think like a lawyer. How does this happen? It happens by a lot of reading. Law students read many court cases. They study the facts. They study the arguments that are made on each side. They also study what decision the judge made in each case and why.

You have to read a lot to ▶
become a lawyer!

◀ Studying past
court cases

In this way, law students learn how to choose which laws are related or connected to the facts of a certain case. They learn how to apply the law. After learning the basics of law in general, some students may choose to focus on intellectual property law.

A law student in ▶
a law library

In the United States, every lawyer needs to pass a state bar exam to become a licensed lawyer. This test covers state and national laws. It is not easy. But no one can be a lawyer without passing the state test where he or she plans to work.

A lot to study

Some people fail the first time. That's OK, though. Even some famous people failed the first time. Those who fail can take it again. But if they fail too many times, they might need to take some more classes.

Not passing the first time

Intellectual property lawyers work with national patent and trademark offices. They must also pass a special exam for patent lawyers. To be an intellectual property lawyer, there is a lot of studying. There are a lot of tests. But those who are determined and focused can get through it.

Books about patent law ▶

◀ More to study

Good intellectual property lawyers need to think and reason logically. They need to ask the right questions and find solutions. Strong speaking and writing skills are important, too. In court, they will need to convince people to believe them. Sometimes, the difference between two patents may be a very small thing. So intellectual property lawyers must pay close attention to details.

◀ An intellectual property lawyer with strong speaking skills

Looking at ▶
small details

Looking to the Future

In today's world, people are always trying to make things bigger, faster, and better. New inventions are being made all the time to solve problems or to make life easier. So, much like all through human history, people will always be coming up with new products and new technologies. Look at smartphones! Every year, a newer, better one comes out.

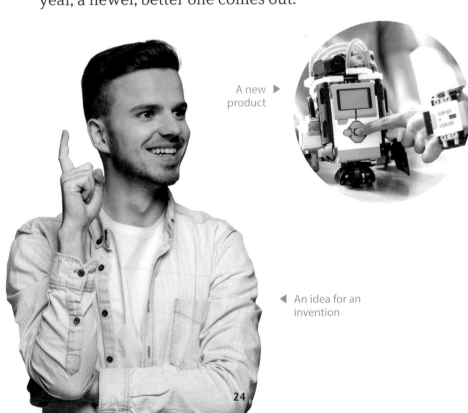

A new ▶ product

◀ An idea for an invention

Getting help from an intellectual property lawyer

In addition, people are always creating songs, writing books, and making art. This means intellectual property lawyers will always be needed. This is good news for future intellectual property lawyers!

Comprehension Questions

1. What is an example of a brand name?
   - (a) Samsung
   - (b) Phone
   - (c) Copyright
   - (d) Patent laws

2. How are written works protected?
   - (a) With a patent
   - (b) Through a trademark
   - (c) By their copyright
   - (d) All of the above

3. What would an intellectual property lawyer NOT do?
   - (a) Apply for patents
   - (b) Write agreements
   - (c) Look for trade secrets
   - (d) Defend patents in court

4. What do law students need to read a lot of?
   - (a) Science books
   - (b) Court cases
   - (c) Newspapers
   - (d) Dictionaries

5. What is NOT a skill an intellectual property lawyer should have?
   - (a) Logical thinking skills
   - (b) The ability to sense danger
   - (c) Attention to detail
   - (d) Good speaking skills

Glossary

- **accuse** (v.) to say that someone has done something wrong or is guilty of something

- **agreement** (n.) a written or spoken promise

- **claim** (v.) to say something as if it is a fact but without showing proof

- **contract** (n.) an official written agreement

- **copyright** (n.) the legal right to publish a piece of writing or perform a song

- **defend** (v.) to protect someone or something from attack

- **intellectual** (adj.) connected with or using a person's ability to think in a logical way and understand things

- **interfere** (v.) to get involved in and try to influence a situation that does not concern you

- **lawsuit** (n.) a claim or complaint against someone that a person or an organization can make in court

- **license** (n.) a document showing legal or official permission

- **logic** (n.) reasonable or sound thinking

- **logo** (n.) a printed design or symbol that a company or an organization uses as its special sign

- **loophole** (n.) a mistake in the way a law, contract, etc. has been written that enables people to legally avoid doing something

- **loss** (n.) an amount or something that has been lost

- **patent** (n.) an official right to be the only person to make or sell a product or an invention; a document that proves this right

- **permission** (n.) the act of allowing someone to do something

- **property** (n.) something that is owned by someone

- **smartphone** (n.) a phone that can connect to the internet and do many of the things a computer can do

- **sue** (v.) to make a claim against someone in court about something that he or she has said or done to harm you

- **trademark** (n.) a name, symbol, or design that a company uses for its products and that cannot be used by anyone else

Notes

Here are some famous intellectual property cases. Readers may enjoy researching these cases to learn more about what intellectual property lawyers do.

Kellogg Co. v. National Biscuit Co. (1938): The National Biscuit Company (Nabisco) invented the cereal "Shredded Wheat" and sued Kellogg for making the same thing. The court ruled that Kellogg had not violated any trademarks because Nabisco's patents had already expired.

Diamond v. Chakrabarty (1980): Chakrabarty created a bacterium that could break down oil and treat oil spills. This was a case to determine whether genetically modified organisms could be patented. The court ruled that a human-made organism could be patented.

A&M Records, Inc. v. Napster, Inc. (2001): A&M Records accused Napster, a file-sharing platform where users could share and download digital music, of infringing on copyrights. This was the first major case to apply copyright laws to file sharing.

Starbucks v. Freddocino (2016): Starbucks filed a lawsuit against Coffee Culture Café for creating a drink, Freddocino, that looked and sounded like Starbucks's Frappuccino, which is trademarked.

List of Books

LEVEL 1

❶ Robotics Engineers

❷ Cyber Security Experts

❸ Medical Scientists

❹ Social Media Managers

❺ Asset Managers

LEVEL 2

❶ Drone Pilots

❷ App Developers

❸ Wearable Technology Creators

❹ Computer Intelligence Engineers

❺ Digital Modelers

LEVEL 3

❶ IoT Marketing Specialists

❷ Space Pilots

❸ Water Harvesters

❹ Genetic Counselors

❺ Data Miners

LEVEL 4

❶ Database Administrators

❷ Nanotechnology Research Scientists

❸ Quantum Computer Scientists

❹ Agricultural Engineers

❺ Intellectual Property Lawyers

"The future of the economy is in STEM. That's where the jobs of tomorrow will be."

James Brown (Executive Director of the STEM Education Coalition in Washington, D.C.)

Data from the US Bureau of Labor Statistics (BLS) support that assertion. Employment in occupations related to STEM—science, technology, engineering, and mathematics—is projected to grow to more than 9 million by 2022 [in the US alone] … Overall, STEM occupations are projected to grow faster than the average for all occupations.

from *STEM 101: Intro to Tomorrow's Jobs* **US Bureau of Labor Statistics**